CONTENTS

WHAT IS A SNAKE?

There's sssssomething mysterious about snakes, so let's clear the air, shall we? Snakes are long, legless reptiles that live almost everywhere in the world. They are carnivores. Some are venomous and some squeeze their prey to death. Some can spit and others can glide. They are also sssssuper important! Snakes balance ecosystems by eating pests that carry disease, like rodents and insects. They are also food for larger predators, such as birds of prey. And scientists use their venom to make medicines that help people. Like them or not, snakes do a whole lot of good for our planet and the people on it!

ANCIENT REPTILE RELATIVES

While snakes may be legless today, their ancestors were not! Dating back almost 100 million years ago to when the dinosaurs roamed the earth, scientists believe modern-day snakes evolved from a four-legged **reptilian** creature. But the biggest and oldest snake fossil on record, dating back 60 million years ago, is Titanoboa. This giant of a reptile measured approximately 50 feet (15 m) long. Its fossil was found in South America.

FANG YOU VERY MUCH

With more than 3,600 species of snakes in the world, there are a lot of differences within the snake family. And when it comes to cool and dangerous snakes, some of those differences can mean life or death.

venomous vs. poisonous

A venomous snake or animal is one that injects venom into its prey in order to kill and eat it. A poisonous snake or animal is when another animal gets sick from eating the poisonous animal.

fangs vs. teeth

Almost all snakes have teeth, but not all snakes have fangs. Only venomous snakes have fangs. The fangs are hollow so the snake can inject venom once it has bitten into its prey.

venom vs. Antivenom

Antivenom is the most effective treatment for deadly snake bites, and it is made out of real snake venom! Much like scientists have created **immunizations** for diseases, scientists use the venom from snakes to create the antivenom through a difficult—and expensive—process.

nonvenomous vs. constrictors

Just because a snake isn't venomous doesn't mean it can't be deadly. **Constrictors**, especially anacondas and pythons, are known for killing large prey, including deer, caiman, and yes, even humans! Most snakes are nonvenomous, meaning they can leave a painful bite but they don't inject venom.

TOP TEN MOST DANGEROUS SNAKES

The fear of snakes, called **herpetophobia**, is one of the most common fears in the world, but the chances of getting bit by a venomous snake—and dying from it—are extremely rare. Human deaths by snake bite happen most often in places where medical help and antivenom are not easily available. This is why the snakes with the deadliest venom may not be the most dangerous. It's also why the official order of the most deadly snakes can change from year to year. Most snakes bite humans because they feel threatened or have been disturbed.

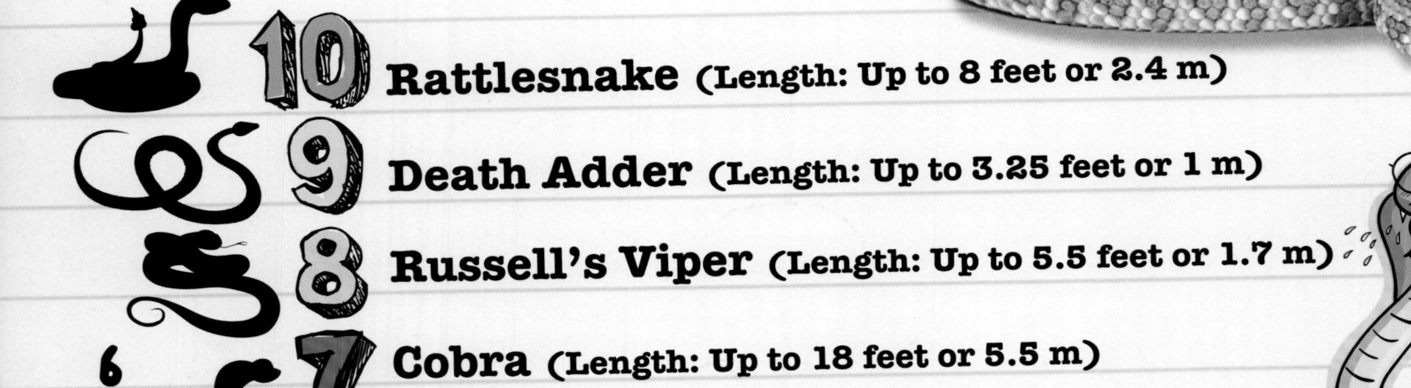

10 **Rattlesnake** (Length: Up to 8 feet or 2.4 m)

9 **Death Adder** (Length: Up to 3.25 feet or 1 m)

8 **Russell's Viper** (Length: Up to 5.5 feet or 1.7 m)

7 **Cobra** (Length: Up to 18 feet or 5.5 m)

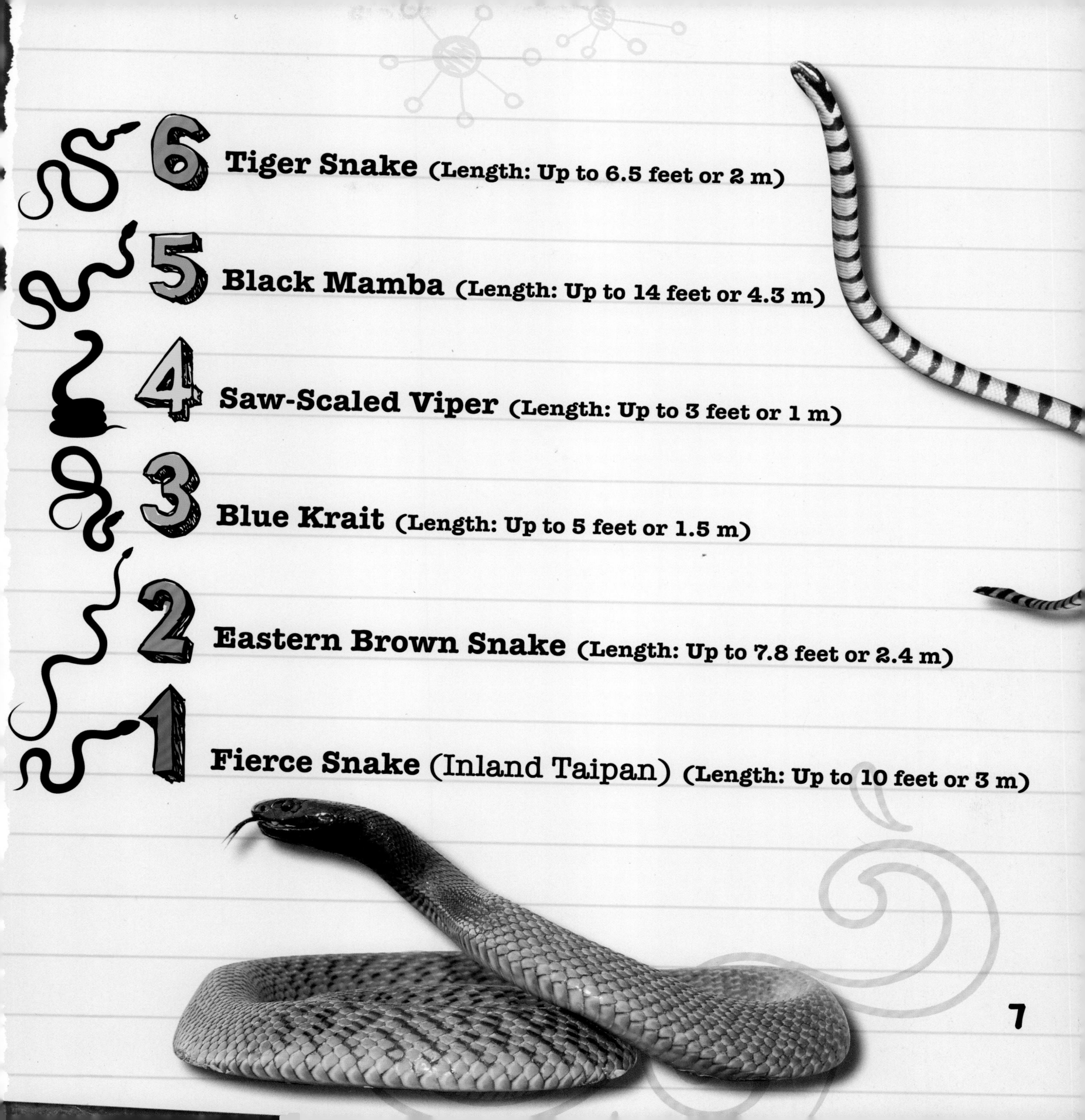

6 Tiger Snake (Length: Up to 6.5 feet or 2 m)

5 Black Mamba (Length: Up to 14 feet or 4.3 m)

4 Saw-Scaled Viper (Length: Up to 3 feet or 1 m)

3 Blue Krait (Length: Up to 5 feet or 1.5 m)

2 Eastern Brown Snake (Length: Up to 7.8 feet or 2.4 m)

1 Fierce Snake (Inland Taipan) (Length: Up to 10 feet or 3 m)

#10 RATTLESNAKE

- Rattlesnakes are in the pit viper family. There are 33 different types of vipers **classified** as rattlesnakes.

- The rattle is made up of dried hollow segments of skin made of **keratin**—the same material human hair and nails are made up of. The snake gets a new row of rattles every time it sheds its skin.

- The most dangerous of the rattlesnakes is the Pacific Rattlesnake. Its venom is more potent than some of the larger and more common species, such as the western diamondback.

- A group of snakes is called a bed, den, or pit, but a group of rattlesnakes is called a **rhumba**.

SNAKE BITE

Some snakes can have more than 200 teeth, but they're not for chewing. These teeth point backward so the prey can't climb back out of the snake's mouth.

SNAKE STATS

Length: Up to 8 feet (2.4 m)
Food: Small mammals and birds. Sometimes other snakes, lizards, and large insects.
Location: Canada, Argentina, North America, and Mexico

#9

DEATH ADDER

- To entice its prey, a death adder "wags" or "whips" its tail.
- Death adders give birth to live young, anywhere from 10 to 30 at a time.
- A puff adder is in the death adder group, and gets its name from the sound it makes when it inflates its body and makes a hissing sound when threatened.

SNAKE BITE

Death adders are related to the cobra, but they look more like vipers, with thick bodies, short tails, and large heads.

Length: Up to 3.25 feet (1 m)
Food: Frogs, young birds, and small mammals
Location: Deserts and rain forests in Australia and New Guinea

#8 RUSSELL'S VIPER

The venom of a Russell's viper is highly **fatal**. It causes so many deaths because it lives mostly on farms where it comes into contact with humans more often than most other snake species.

A Russell's viper has a distinctive pattern, three rows of reddish-brown spots that are outlined in black and white.

The Russell's viper is named after the 18th-century scientist, Patrick Russell, who recorded some of the first information about snakes in India.

SNAKE BITE

A snake loses its fangs every six to 10 weeks. New ones grow in their place.

SNAKE STATS

Length: Up to 5.5 feet (1.7 m)

Food: Small mammals, birds, lizards, and insects

Location: India and Taiwan

#1
COBRA

- A cobra's distinctive "hood" is formed when it spreads a part of its body called the **neck ribs**, muscles that spread out the loose skin around its head.

- The cobra's biggest enemy, aside from humans, is the **mongoose**, a small carnivorous mammal.

- In some countries, snake charmers capture cobras for display, removing their fangs and sewing their mouths shut. A cobra can't actually hear the music; it responds to the movement and vibrations.

SNAKE BITE

The king cobra is the largest snake in the cobra family, measuring up to 18 feet (5.5 m) long!

Length: Up to 18 feet (5.5 m)

Food: Other snakes

Location: India, Southeast Asia, Philippines, and Indonesia

#6 TIGER SNAKE

- Tiger snakes are very shy but when threatened, they will inflate and deflate their body, making a hissing sound.
- They are a part of the cobra family.
- The tiger snake's venom contains toxins that **clot** a person's blood and paralyzes their **nervous system**.
- Tiger snakes are born with stripes, but as they grow older, those stripes fade in many of the species.

SNAKE BITE

A **herpetologist** is a scientist who studies reptiles and amphibians.

SNAKE STATS

Length: Up to 6.5 feet (2 m)
Food: Frogs, lizards, birds, small mammals, and fish
Location: Australia and its nearby islands

#5 BLACK MAMBA

- The black mamba's scales are olive green, but its mouth is black, which is where it gets its name.
- It prefers to sleep in termite mounds or hollowed-out trees.
- When a black mamba strikes, it will bite its victim again and again. Two drops of its venom can kill most humans if untreated.

SNAKE STATS

Length: Up to 14 feet (4.3 m)
Food: Small mammals and birds
Location: Africa

SNAKE BITE

Black mambas are one of the fastest snakes—clocking in up to 12 miles per hour (19 kph)!

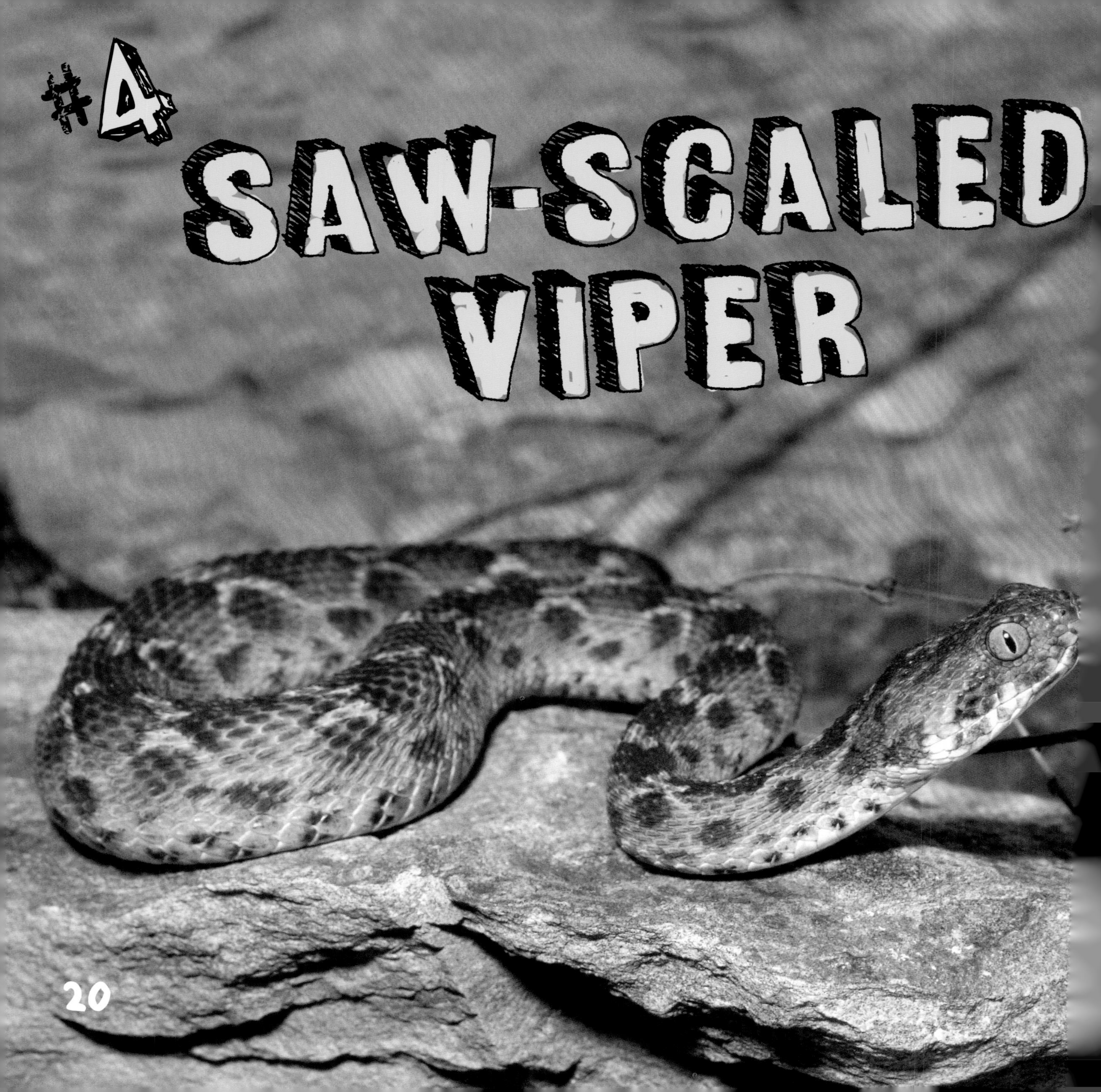

#4 SAW-SCALED VIPER

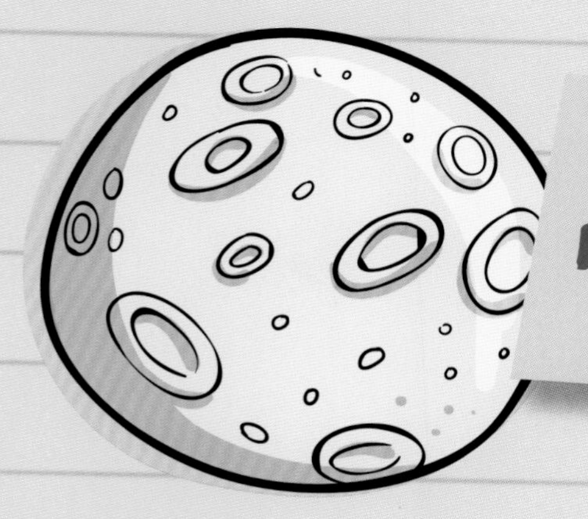

SNAKE BITE

Saw-scaled vipers are **nocturnal**, hunting prey at **dusk**.

- Saw-scaled vipers are named for their **serrated** scales, which are found right below their head.
- They are sidewinders, which means they move their bodies in a sideways loop, with no more than two points touching the ground at the same time. This keeps their bodies cooler as they move across the desert sand.
- The saw-scaled viper, an **aggressive** snake, causes more deaths in its region than other snakes because of its lethal venom and the lack of access to antivenom.

SNAKE STATS

Length: Up to 3 feet (1 m)
Food: Mammals, birds, snakes, lizards, and insects
Location: Africa and Asia

#3 BLUE KRAIT

- Despite its name, the blue krait is not actually blue! It is striped with yellow, brown, black, or bluish-black bands while its belly is white.
- They are highly venomous but rarely strike a human unless they are stepped on or threatened.
- The blue krait's venom is a **neurotoxin**, meaning the venom shuts down a person's nervous system, including the lungs, heart, and brain. Most deaths are caused by suffocation.

SNAKE BITE

Also known as the Malayan krait, the blue krait is a member of the **elapid** family, which are snakes that have hollow fangs that do not retract.

SNAKE STATS

Length: Up to 5 feet (1.5 m)

Food: Snakes, lizards, and small mammals

Location: Asia and Indonesia

#2 EASTERN BROWN SNAKE

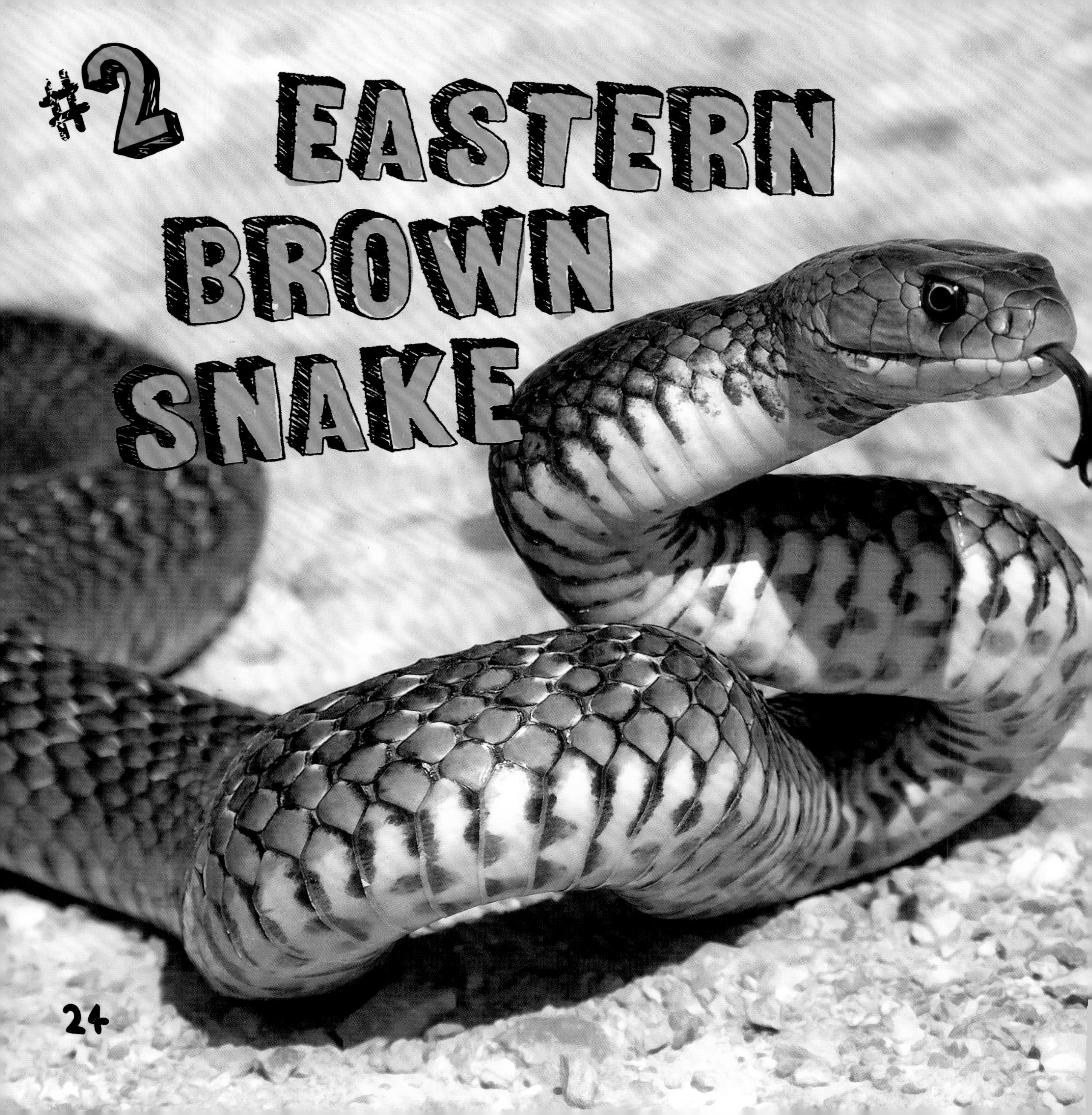

- The eastern brown snake is the second deadliest snake on Earth. More people in Australia have died from its bite than from any other snake.

- Because it preys on small rodents, such as house mice, they are mostly found on farms and near homes.

- Male eastern brown snakes compete for females during mating season in a **combat ritual**, which looks like twisted rope.

SNAKE BITE

A group of snake eggs is called a **clutch**. Once they hatch, **juvenile** snakes are on their own.

SNAKE STATS

Length: Up to 7.8 feet (2.4 m)
Food: Rodents
Location: Australia and New Guinea

#1 FIERCE SNAKE

(Inland Taipan)

The fierce snake has the most toxic venom of all snake species. One bite has enough venom to kill about 80 humans.

Even though its venom is the deadliest, few deaths are recorded because it lives in remote places far away from humans, which means bites are deadly but rare.

Unlike most venomous snakes that deliver their venom in one strong bite, the fierce snake bites repeatedly.

SNAKE STATS

Length: Up to 10 feet (3 m)

Food: Small mammals

Location: Australia

SNAKE BITE

The fierce snake changes the color of its skin according to the season. They are lighter during the summer and darker during the winter. This allows the snake to absorb more warmth in the cooler months.

Other cool and dangerous snakes

MOZAMBIQUE SPITTING COBRA

- The Mozambique spitting cobra can eject its venom up to 8 feet (2.4 m) away. It aims its venom at its prey's eyes, which can cause blindness.
- To escape a predator, Mozambique spitting cobras have been known to "play dead."
- Its venom is extra potent because it is a combination of a neurotoxin and **cytotoxin**, meaning it attacks the nervous system and soft tissue.

SNAKE STATS

Length: Up to 5 feet (1.5 m)

Food: Amphibians, snakes, birds, eggs, small mammals, and insects

Location: Africa

SNAKE BITE

When threatened, this snake will rear up its head and body to two-thirds its length.

LANGAHA SNAKE

- Also known as the leaf-nosed snake, this snake is an **arboreal** species, meaning it lives in the trees of Madagascar forests. It is named for its leaf-shaped nose—a pointed "horn" on its snout.
- Their venom is painful but not deadly to humans.
- Males are more colorful than females. Male and female have different-shaped snouts.
- The fangs of a leaf-nosed snake are found in the back of the upper jaw. It chews on its prey to make sure the venom gets in.

SNAKE STATS

Length: 3 feet (1 m)
Food: Lizards
Location: Madagascar

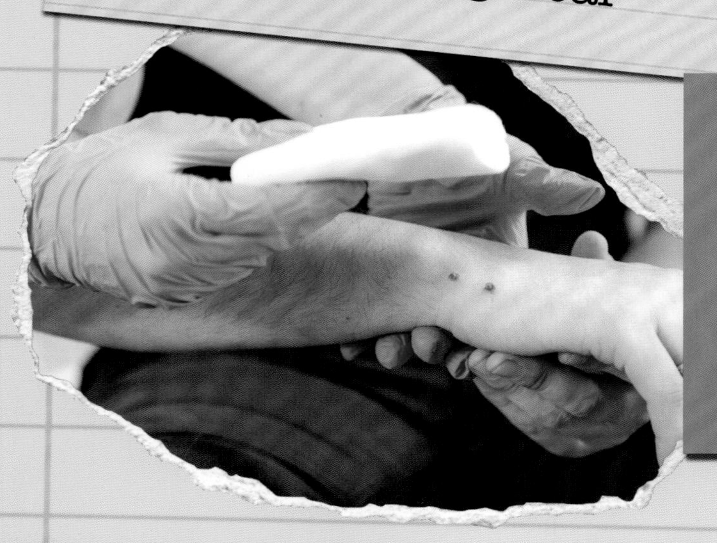

SNAKE BITE

If you get bitten by a snake, apply pressure to the puncture site. Keep as still as possible and call 911.

ATHERIS VIPER

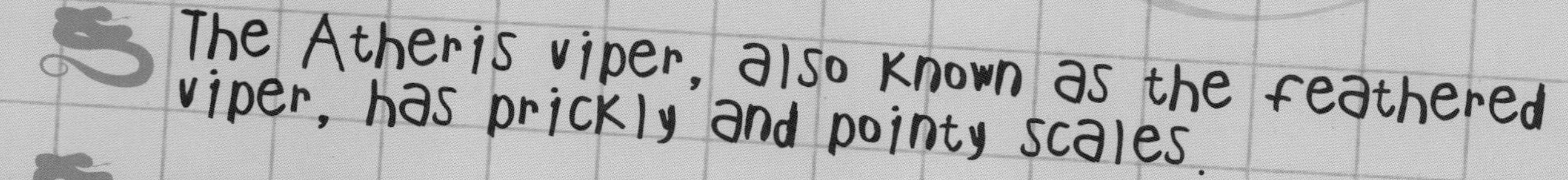

- The Atheris viper, also known as the feathered viper, has prickly and pointy scales.
- There is no known antivenom to this snake's venom.
- Atheris vipers are **ovoviviparous**, which means they give birth to live young.
- Their venom is hemotoxic, meaning it causes pain, swelling, and blood clotting issues. This can cause severe internal bleeding in a human.

SNAKE STATS

Length: 16 inches (40 cm)
Food: Small amphibians, lizards, small rodents, birds, and other snakes
Location: Africa

SNAKE BITE

Atheris vipers often attack their prey by hanging upside down from a tree and **ambushing** it from above.

BURROWING ASP

- The burrowing asp can bite its prey without opening its mouth. Its fangs are also flexible and can move independently.

- Like its name suggests, this species burrows tunnels and lives underground.

- Burrowing asps stab their prey sideways. Some scientists believe this is because most hunting happens in small burrows underground with little room to open their jaws.

SNAKE STATS

Length: Up to 2.5 feet (76 cm)

Food: Small rodents, birds, lizards, and insects

Location: Africa and Middle East

SNAKE BITE

Its venom can cause **tissue necrosis** or the deterioration of its prey's flesh if left untreated.

SEA SNAKE

- Sea snakes, such as the Belcher's sea snake, have venom that is six times more potent than the deadliest cobra.
- There have been sixty-two species of sea snakes recorded—and most of them are extremely venomous.
- All sea snakes have paddle-like tails, which make them look like eels. It also makes them very clumsy on land.
- Very few deaths happen from sea snake bites because they rarely inject venom into their prey. This is called a **dry bite**.

SNAKE BITE

Most sea snakes are fully adapted to living a **marine life**, with nostrils that are covered to keep water out. However, they must still surface for air.

SNAKE STATS

Length: Up to 9 feet (3 m)
Food: Small fish and shellfish
Location: Indian and Pacific Oceans

ANACONDA

There are currently four species of anacondas, but the most well-known is the green anaconda, which is the largest snake in the world by weight and one of the longest.

A green anaconda is a constrictor, which means it kills its prey by wrapping itself around it, squeezing it tighter until the prey can't breathe or it is crushed.

It unhinges its jaw to eat its prey whole. After a big meal, an anaconda can go weeks without eating again!

SNAKE STATS

Length: Up to 29 feet (8.8 m)
Food: Fish, birds, wild pigs, capybaras, caiman, and deer
Location: South America

SNAKE BITE

Some green anacondas have been known to capture and eat jaguars!

FLYING SNAKES

- Much like flying squirrels, flying snakes don't really fly, but they sure can glide! These snakes can glide up to 330 feet (100 m).
- Flying snakes climb trees using special scales on their bellies. When they want to glide, they dangle themselves off a branch, forming a "J" shape and then push their bodies forward, stretching out their ribs to make a winglike parachute.
- They are mildly venomous but spend most of their time up in trees and, therefore, don't interact much with humans.

SNAKE BITE

The smaller species of flying snakes are better gliders than the larger ones.

SNAKE STATS

Length: Up to 4 feet (1.2 m)

Food: Lizards, rodents, frogs, birds, and bats

Location: Asia

RETICULATED PYTHON

The **reticulated** python is one of the longest snakes in the world. They are constrictors, like the anacondas, who ambush their prey.

They are excellent swimmers and have been known to **colonize** small islands far out in the sea.

Reticulated pythons are often kept as pets, but there are official reports that they have been known to kill humans. While not common, it is important to be safe!

SNAKE BITE

Reticulated means "net-like," which describes the intricate skin pattern of these pythons.

SNAKE STATS

Length: Up to 22.8 feet (6.9 m)
Food: Mammals, birds, primates, and pigs
Location: Asia

HORNED VIPERS

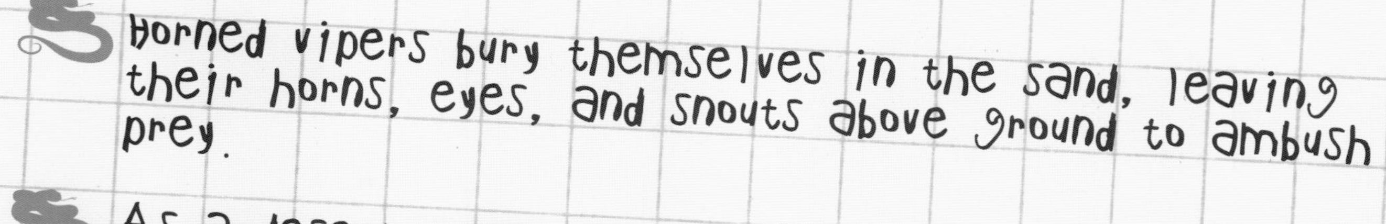

- Horned vipers bury themselves in the sand, leaving their horns, eyes, and snouts above ground to ambush prey.
- As a desert snake, the horned viper "sleeps" during the summer, much like bears hibernate in the winter. It's called **aestivation**.
- The head of a horned viper is covered in rough scales with ridges. Most have a spine-like horn over each eye—but not all! Some snakes born from the same litter will have horns and others will not.

SNAKE STATS

Length: Up to 20 inches (50 cm)
Food: Rodents, birds, and lizards
Location: Africa

SNAKE BITE

Horned vipers can disappear quickly by using their rough scales in a rocking motion to help them "sink" into the sand.

CORAL SNAKES

- Coral snakes are known for their bright colors and patterns.
- There are 85 species of coral snakes in the world, and they can be divided into two subgroups: New World coral snakes are found in North America, and Old World coral snakes are found in Asia.
- The king snake is a species that resembles the North American coral snake. It has the same colors but a different pattern!
- Coral snake venom is a neurotoxin, which paralyzes its prey's breathing muscles, causing suffocation.

SNAKE STATS

Length: Up to 5 feet (1.6 m)
Food: Small snakes, lizards, frogs, small birds, and rodents
Location: North America and Asia

SNAKE BITE

Remember this rhyme to spot the difference between a coral snake and a king snake: "Red touch black, safe for Jack. Red touches yellow, kills a fellow."

GLOSSARY

Aestivation—hibernating in the summer months instead of the winter

Aggressive—attacking without warning or reason

Ambushing—attacking prey by surprise

Antivenom—the medicine used to neutralize a deadly snake bite

Arboreal—using colors, materials, or light to disguise oneself as protection

Carnivore—an animal or plant that eats meat

Classified—when species are divided into groups with similar characteristics

Clot—when blood cells form lumps and stop flowing

Clutch—a nest of snake eggs

Colonize—to move to a new place and make it home

Combat ritual —a traditional fight between two male snakes for the attention of a female

Constrictors—a snake that wraps its body around its prey and squeezes it to death

Cytotoxin—toxic to human tissue

Dry bite—when a venomous snake bites its prey but does not inject venom

Dusk—the part of the day where it starts to get dark

Ecosystems—a community of living and nonliving parts of an environment that work together

Elapid—the family of venomous snakes that have hollow fangs

Fatal—causing death